Frozen Snakes and
Dinosaur Bones

Margery Facklam

FROZEN SNAKES AND DINOSAUR BONES

Exploring a Natural History Museum

Illustrated with photographs

New York and London
Harcourt Brace Jovanovich

FRONTISPIECE: Putting together a dinosaur skeleton is like working on a giant puzzle. In order to make sure she has all the bones in the right place, the scientist puts each bone in place on the picture before she strings them together.

PHOTO CREDITS

Courtesy of the American Museum of Natural History: pp. 2, 31, 45, 54, 73, 75, 79, 84, 86.
Buffalo Museum of Science: frontispiece, pp. 9, 12, 13, 15, 26, 39, 44, 56, 57, 64, 65, 69, 81, 82, 92, 102, 103.
Ward's Natural Science Establishment, Inc., Rochester, N.Y.: pp. 52, 55.
New York State Museum: pp. 60, 67, 68.
Margery Facklam: pp. 21, 23, 24, 100.

Copyright © 1976 by Margery Facklam

All rights reserved. No part of this publication may be reproduced or transmitted in any form or by any means, electronic or mechanical, including photocopy, recording, or any information storage and retrieval system, without permission in writing from the publisher.

Printed in the United States of America

First edition

B C D E F G H I J K

Library of Congress Cataloging in Publication Data

Facklam, Margery.
Frozen snakes and dinosaur bones.

SUMMARY: Takes the reader behind the scenes in a natural history museum.

1. Natural history museums—Juvenile literature.
[1. Natural history museums] I. Title.
QH70.A1F3   500.9'074   75-41394
ISBN 0-15-230275-1

To Howard

Contents

1 What If You Made a Museum? 3
2 How Do You Make a Museum? 7
3 What Good Are Ten Skunks? 11
4 Ditchdiggers with Notebooks 17
5 Flat Plants 25
6 Six-legged Science 29
7 On Top of the Museum 37
8 Skeletons and Skins 43
9 Skinning the Cat 51
10 Frozen Snakes and Spiders 59
11 Putting It All Together 63
12 Gorillas Are Great 71
13 Dinosaurs 77
14 Mastodon Mystery 87

15	Protecting the Treasure	95
16	Museum Kids	99
17	What's Next?	105
	More Books to Read to Find Out About Museums and the Things They Collect	109
	Index	111

Frozen Snakes and Dinosaur Bones

On December 22, 1877, a new building opened at the American Museum of Natural History in New York City.

1

What If You Made a Museum?

What if someone gave you a huge empty building and said, "Make a museum."

"What for?" you might ask. "Who needs a museum?"

Everyone needs a museum. Museums are for saving things . . . things such as rocks and dinosaur bones and birds and pottery.

Museums save things because scientists and other people want to study them. Learning about stars or birds or rocks can help us live better lives.

"How's that?" you might wonder. While it's true that each of us need not know about birds, it's good someone does. People who built the first aircraft learned about flight by studying the way air moves around birds' wings as they fly. Men and women who study birds can tell us how to get rid of mosquitoes without using poison. They know what kinds of bushes to plant in a yard so that birds that eat mosquitoes will want to live there.

And someone has to study rocks in order to tell people where to build houses that won't get wrecked by earthquakes. Someone has to know about rocks in order to find out where to drill for oil or to discover uranium for nuclear reactors or diamonds for drills or sulfur for matches.

Studying stars is important for space flights. And collecting things from dead Egyptians or ancient Indians helps us learn about people who lived differently than we do now. Sometimes we learn that they had good ideas. Even though they didn't call it ecology, the Indians knew a lot about living with the land. They knew that they shared the land with plants and animals, so they didn't go around chopping down too many trees or killing more animals than they could eat.

When we find out how early people made their tools or what kind of religion they had, we can begin to understand them. And then we can better understand ourselves and the things we do.

Sometimes we collect objects, like ancient pottery, just because we are curious and we like to see things that will never be made again.

It *is* important to collect objects.

"O.K.," you might finally say. "It's important to have collections. But how do I start to make my museum?"

You could begin with an ad in the newspaper:

> Wanted: Objects for a new museum

You might be a guest on a television interview program where you could tell people about your museum and ask them to send things.

Then you could sit back in your director's chair and wait for your museum to grow. And it would grow.

People would send you shrunken heads and arrowheads, fossils and falcons, minerals and masks, bird nests and fishing nets, weeds and wigwams, pieces of pottery and parts of peace pipes.

The zoo director might call one day and say, "Our orangutan died. Do you want him?"

A woman might phone and say, "We have an empty wasps' nest. Can you use it at your new museum?"

Or a man who had collected birds' eggs for 55 years would walk in lugging huge boxes. He would say, "I have 803 eggs. My wife says I have to get them out of the house. I'm giving them to the museum."

And so your museum would grow until one Monday morning you would walk into your big building and it wouldn't be empty anymore. You would trip over boxes and bones and beetles.

The place would look like an enormous attic, crammed full of thousands of things. You wouldn't be able to remember whether the sharks' teeth were in the box under the birds' eggs or next to the wasps' nest. You would not remember who had given what.

Right then you might holler, "Hey, wait a minute. This isn't a museum."

Only no one would hear you because no one works in your museum.

2

How Do You Make a Museum?

In 1846, the Smithsonian Institution in Washington, D.C., had a big problem. The newspapers were calling it the "nation's attic."

It was a new museum, just beginning. People everywhere were bringing things and mailing things. When the Army or the Navy returned from some far part of the growing United States or some foreign country, they brought back animals and rocks and weapons and costumes of the natives. They sent all these things to the new museum.

At first the museum people put things on shelves so visitors could look at them and say, "Wow, look at that. Isn't that weird?"

Museums long ago were called "cabinets of curiosity" because that's just what they were, rows and rows of cabinets full of curious things . . . two-headed calves, potatoes shaped like ducks, stuffed monkeys, and shiny shells.

Museum workers were not happy. They thought that there must be a better way to show people all the strange and beautiful things in the world.

Walking around your museum-that-isn't-a-museum, tripping over boxes, you'd have the same problem the Smithsonian people had. You'd wonder how you could get the mess in order. Whom should you hire to help?

And then, suddenly, you would know what to do. You would get someone to make a list of all the things and put a number on each one. Then you would know exactly what you have.

The person you would hire to get things in order would be called the registrar. The registrar keeps track of everything in the museum. He registers things.

First he might pick up a shrunken head and put a number on it with the kind of ink that doesn't wash off. Then he'd put the same number on a card and type:

>   1. Shrunken head from New Guinea
>   Gift from Mrs. Bertha Barclay,
>   May 31, 1974

A dead bat might be the next thing in your pile of objects. The registrar would tie a linen tag onto the bat's leg and put the number 2 on the tag. The card for the bat might say:

>   2. Bat from Mr. Issac Badger's barn,
>   Snyder, N.Y.
>   Caught October 6, 1975

And number 3 might be many small pieces of one object, such as a broken bowl. Then the registrar would use the ink to mark each broken piece with a number. The card would be marked with all the numbers, like this:

3. Twelve pieces of pottery from Arizona
Collected by Karl Kramer,
February 21, 1976
Numbers 3a, 3b, 3c, 3d, 3e, 3f, 3g, 3h, 3i, 3j, 3k, 3l

When the registrar had finished numbering all the objects piled in your new museum, you could again lean back in your director's chair, satisfied that things were in order. Everything would have a number and a card, just like the books in the library.

The registrar would interrupt your contentment. "The bat is on a shelf next to the shrunken head, and the rocks are in a cupboard with the insects. The Indian baskets are in a box with the sea urchins. Everything has a number, but the

Two anthropologists check the numbers the registrar listed for some artifacts.

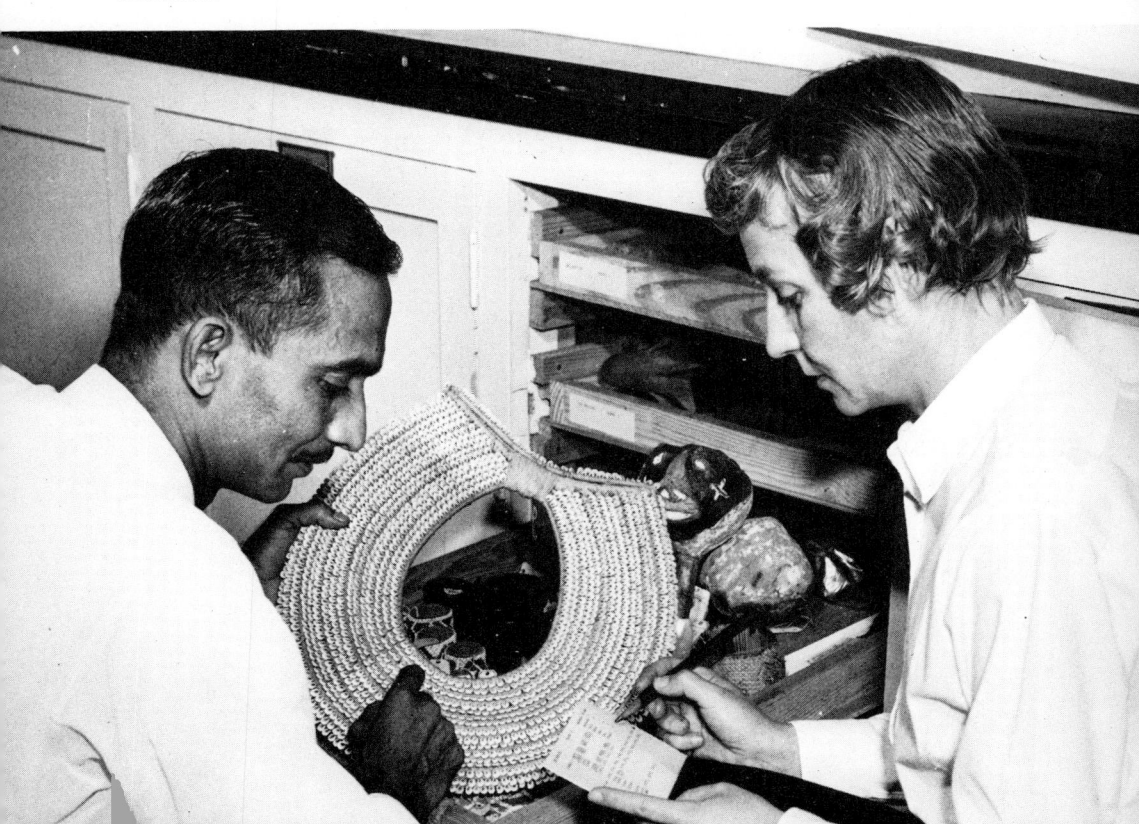

place is still a mess. And two wooden spears are broken."

Then you would realize that a director and a registrar do not make a museum. You would have to hire people to take care of the collections.

You would hire one person to take care of the rocks, another to take care of the animals, and another to take care of the shrunken heads and Indian baskets. These people will be called curators.

Curators are people who take care of collections. The curator of zoology studies and takes care of animals. The curator of botany studies and takes care of plants. The curator of geology studies and takes care of rocks and fossils.

You can have as many curators as there are things to study. You can have a curator for just about every kind of science there is. Or, if yours is a small museum, you might have only one.

The curators also put labels on things. They always use a scientific name first, like the fancy one *Erethizon dorsatus*. Then they add the common name, which would be the Eastern or American porcupine. Scientific names are a common language. They are learned by scientists in every country, no matter what language they speak.

A curator looking at an object in a German museum or an American museum or a Japanese museum can read the labels because they are written with the same scientific name.

Curators also repair broken objects in their collections. They store things carefully so that nothing will be hurt by too much heat or light or insects or dampness.

Curators are the keepers of the treasures of the earth, things we see every day and things that will never be on this earth again.

3

What Good Are Ten Skunks?

One day your curator of geology might walk into your office and say, "I have to get some more rocks."

"More rocks? Good grief, you have drawers full of rocks," you'd tell him.

Just then the curator of zoology might poke her head into the doorway and say, "I have to go out, too. A lady wants me to pick up a dead skunk in her yard."

"Another skunk," you'd yell. "Don't we have enough? The forest ranger brought in two yesterday that had been killed on the road. What can you do with so many skunks?"

"Let me ask you something," the curator of zoology might say. "What if you were a person from another planet and you asked me what the average earth man looked like? And what if I told you that the next man off an elevator would be the average earth man? Would that be true?"

You might think a minute, but then you'd say, "Hey, no. The first man off the elevator might be bald and have one wooden leg and be four feet tall. He wouldn't be average."

"That's right," the curator would agree. "And that's just why I need more than one skunk. Maybe one of my dead skunks has no tail, or another doesn't have a white stripe, or another might be a tiny runt of the litter. I can't tell much about an average skunk unless I have many of them to compare."

Then your curator of geology would explain how he needs all different kinds of rocks—from mountains, mines, oceans, deserts, quarries, and even the moon.

He would tell you that geologists learn how to read the history of the earth in layers of rocks. They can discover where great oceans once washed over land and where rivers of ice once flowed over forests. By studying rocks, geologists can find out where volcanoes will erupt, where earthquakes might occur, or where mountains might move. They can tell

The scientists who study rocks often work with scientists who study living things. When geologists temporarily changed the flow of water over Niagara Falls so they could study the rocks underneath, biologists went along to collect the plants and animals that live in the water there.

Curators are always trying to add to their collections. These arrow points were all collected at the same place, but they are not all alike. They may have been made by different people or traded from another tribe.

where to find all the metals and elements we need for fuel and for manufacturing things from matches to space ships.

"All right," you'd say to your curators. "Go and get skunks and rocks and whatever else you need for your collections."

A curator is a scientist, and a scientist is a detective. He finds clues and pieces the facts together to explain things.

A scientist can't learn much from one spearhead. Maybe the caveman who made it was clumsy and made a dull spearhead that didn't work very well. Maybe all the other cavemen who lived at that time made beautiful, sharp spearheads. From the dull spearhead the scientist might think that all cavemen were clumsy and didn't know much about working with stone. But he really wouldn't know that until he had lots and lots of spearheads from the same place to compare.

If you found a bunch of the same kind of pottery bowls buried at the place where there had been an ancient village, you might well be able to say that this is the kind of bowl everyone used for cooking.

But if you found only one pot, you could only guess. One person had a bowl. Maybe he was a rich man. Maybe he was the king and only kings had bowls. Or maybe this group of people didn't know how to make bowls at all and traded with another tribe.

Because they need more than one clue, curators at museums are always trying to add to their collections. They have to decide what is important to collect. Then they have to find places to get what they need. Sometimes they buy objects. Sometimes they trade. If a geologist has dozens of quartz crystals because they are found near his museum, he might trade them for some fossils found near another museum.

Often curators go on trips, called expeditions, to find things for their collections. The farthest, most expensive expedition ever was the trip to the moon, where the astronauts, who were also geologists, collected moon rocks. But some expeditions are only one-day trips to nearby fields to find insects.

Some of your curators will be away from the museum for many months. They might be studying bird migrations in Mexico or digging dinosaur bones in Wyoming or tracking grizzly bears in the Rockies. Then, when they return from their trips, they will spend many more months organizing the collection and information found on their expeditions.

Some curators go on digging expeditions to find fossils.

Your geologist would come back from his rock hunting with lots of rocks and a notebook that tells exactly where each rock was found. The most important word to a geologist is "where." In the museum, his biggest problem is finding places for all his samples. Rock collections cause few storage problems other than space. They are seldom bothered by insects or dampness or heat.

Each of the rocks your curator of geology collected is marked with a dab of white paint. The painted spot makes a smooth surface for a name or number to be printed in ink. Just as in all other collections, each number on an object matches a number on a card in a file that also tells where it was found and when and by whom.

Curators teach people about their collections. But the collections themselves are what make museums different from schools where people teach and study, too. Museums are museums because they save things in an orderly way.

4

Ditchdiggers with Notebooks

The person you hire to take care of your arrowheads and shrunken heads should be very nosy. That person should never want to stop asking questions about people who lived long ago, questions about what they ate and how they hunted and how they grew food and what kind of houses they lived in. That person would be your curator of anthropology.

The word anthropology is easy when you take it apart. The first half, *anthrop*, means man, and the second part, *logy*, means study. So an anthropologist studies man.

Your curator of anthropology might be a special kind of anthropologist called an archeologist, a person who learns about prehistory. We have no written records from the time we call prehistory. As far as we know, there were not any books or magazines or slides or movies. But an archeologist learns from the things people left behind: pottery, tools, weapons, and even garbage. An archeologist can tell just from looking at the marks or the shine on a tool whether it was used for cutting meat or harvesting grain. He can look at animal bones under a microscope and tell if the animals were wild ones killed for food or tame ones raised by farmers.

All the bits and pieces of things are called artifacts. An archeologist collects artifacts for his museum.

What if thousands of years from now something happened on earth so that our cities were buried and all the books and papers were destroyed? And what if some scientists from another planet started digging through the rubble to find artifacts from the people who had once lived on earth? What if they found some television tubes, false teeth, a Barbie doll, and some ice skates? What if they found a lawn sprinkler, a broken Ferris wheel, and an eggbeater? What if they found an aluminum soda-pop can, a yoyo, and a plastic dinosaur model?

Do you think they could look at these things and know, really know, how we lived? If they figured out what a Ferris wheel was, would they know what it was like to ride around and around on a hot summer afternoon and drink some cold soda-pop afterward? Would they know what it was like to watch television or run under the sprinkler? Would they think a Barbie doll was an idol we worshipped or that ice skates were strange knives people wore on their feet to cut noodles?

Of course they couldn't really know. And so we don't really know all about people who lived in prehistory. We know what they ate and wore and worked with. But we can only imagine what it was like when someone first learned to make fire or invented a needle to sew clothes together.

We don't know what people in prehistoric times called themselves or what language they spoke. Often we just name them after the place they were found, such as the cavemen or the lake dwellers. Or we name them after a tool or weapon they used, such as the Battle-Ax people.

When museums were beginning, the collectors they sent out brought back jewels and golden statues and spectacular treasures. They didn't pay any attention to little things like pieces of broken pottery or chips of stone or bones and beads. They missed some of the most valuable clues.

Then some archeologists began to wonder about the people who made the treasures found in tombs and pyramids. What were they like? Did their children have toys? What kinds of tools did they use? And so the archeologists went looking for the missing clues.

A place where an archeologist digs is called a site. At first, finding sites was just plain luck. A highway being cut through a hill would open up an ancient village site or a trade center. A foundation being dug for a building would show artifacts as the steam shovel scooped out dirt.

In Africa there is a famous site called Olduvai Gorge that was opened by wind and rain that washed topsoil away. It apparently was the site of an ancient lake where early human beings fished and worked. Dr. Louis Leakey, a famous archeologist, believed that the earliest humans yet discovered lived there because he found skulls of these people along with bones of the animals they ate and the tools they used.

Even explosions sometimes help uncover sites. During World War II, artillery fire in Italy exposed caves loaded with the remains of an early group of people.

Archeologists didn't want to depend on luck all the time, so they learned to find sites. They listened to legends and folk stories about buried cities and old ruins. They read very old journals of the first Jesuit missionaries who traveled among the early Indians in North America. There would be pages telling that when you walk inland two days from the lake, the village would be among a stand of pines. Of course, it might take years to figure how far two days inland was and in what direction. But at least they knew a village had stood and there was something to look for.

Walking over plowed land, an archeologist can tell if there is something to dig for. He feels the soil for a greasiness left from ancient cooking sites or thrown-out dish water. He looks for chips of stone and pieces of bone.

He looks for mounds or small hills in an otherwise flat plain. A mound builds up over centuries of village living. A house built of mud and wood begins to fall apart from rain and the normal wear of a family. The next family might use the roof beams of the worn-out house and build a new house on top of the old one. After a long time the ground builds up. Even in our own cities mounds are growing. Old buildings are torn down and new ones put up, and each time some of the old is left underneath. Every thousand people in this country throw away about a ton of rubbish every day. Some of it piles up, and mounds are built.

After an archeologist picks a site, he and his team dig a trench. A trench is a long, narrow ditch. It is a way to find a sample of what is there without digging up the whole field.

If, in the trench, the archeologists find evidence of once-lived life, a bead, a bone, a shard (which is a piece of broken pottery), they really dig in. It is dirty, hard work.

They don't just dig anywhere, making holes all over the site. A very careful plan is followed. The ground is marked off like a giant piece of graph paper, in squares. Each square is numbered. Then if they find an arrowhead, they mark in their notebooks that it was in square B, or wherever they found it. Knowing exactly where an artifact was found is important. A group of arrowheads with a small pile of flint in one square might tell the diggers that this was the place in the longhouse where someone worked on making weapons.

Dark, charred spots on the ground are marked in the notebook, too. They might be the place fire was built, or where posts were placed to build a house. Every spot of color different from the soil is carefully marked in the notebook.

Archeologists dug carefully in a field where they guessed an old fort had been, and they found the outlines of a building.

At the museum the curators of archeology may work for years sorting out the artifacts from one site, repairing them and studying them. The archeologists need the help of other curators, too. They ask the geologists to tell them how old the rock and soil layers are at the site, and they ask the botanists to look at pollen under the microscope to tell them what kind of plants lived there.

Archeologists are only beginning to understand the people who helped make our civilization what it is, our ancient ancestors. Much of their story is still hidden in the earth and under the sea.

Artifacts from a dig might look like a pile of junk, but when each piece is labeled and sorted, an archeologist can begin to learn about life long ago.

5

Flat Plants

The phone rang at the Poison Control Center of a big city. The person who answered asked, "What kind of berries did your daughter eat?"

"I don't know," the caller almost screamed. "But I can tell you what they looked like. They grow in our neighbor's yard. What should I do?"

"Quickly, bring your little girl to the hospital and bring some of the berries with you," said the woman at the Poison Control Center.

When the worried father hung up, the director of the center made a call to the museum. She asked to speak to the curator of botany, the person who studies plants. A few minutes later the curator of botany grabbed his coat and hurried to the hospital to tell the doctors what kind of berries the child had eaten so they could tell what kind of treatment to use.

That same day the curator also answered phone calls from people asking how to get rid of poison ivy or whether it was safe to eat the leaves of the wild grape. They wanted to know how to make tea from hemlock needles and whether you should put tulip bulbs in the ground upside down or rightside up. Three letters in the mail that day had parts of plants in them that people wanted identified.

The botanist spends a lot of time picking flowers and telling other people not to. He tries to teach people which flowers are protected by law so they won't become extinct as many animals have.

Have you ever collected leaves and pressed them in a phone book? If you glued your leaves to stiff paper, you had the beginning of an herbarium, which is what a flat plant collection is called. These flat plants are important.

The botanist puts his collection of flattened plants on sheets of paper, which are called herbarium specimens.

Flat Plants

When a botanist makes an herbarium, he dries and flattens the plants in a press. Then he mounts them on heavy paper and puts a label in the lower right-hand corner. The label tells who collected it, when, and where. It tells the plant's scientific name and common name.

If the plants aren't labeled, they might as well be thrown away. Without data, collections are practically useless. Data is an important word to a scientist. It means facts. It doesn't do much good to have a bunch of plants if you don't know where they came from or what they are.

In your museum, your curator of botany would begin to collect samples of all the plants growing in your county or state. It would take a long time, but someday scientists from other museums could visit your museum to compare plants.

Before World War II, there was a botanist in Germany who had a large herbarium of plants from his country. One day, a botanist in America asked if she could borrow his herbarium to study. The German botanist was happy to send his flat plants to her. Now the American botanist was a woman who loved to collect things. While she had the German plants, she snipped off a small piece of each one, not very much, but just a leaf here and there. She kept these plant pieces in her collection and returned the herbarium to Germany.

During World War II cities were bombed and of course museums were damaged. The German botanist's entire herbarium was destroyed. After the war, the German scientists were grateful that the American botanist had kept even so small a piece of the lost German plants, for those pieces can be used for study and compared to plants now growing in Germany.

About once a year the curator of botany does his housekeeping. He fumigates the plants by putting mothballs or other chemicals in the drawers holding the herbarium to kill the insects who have decided to snack on the dried leaves. There are little black beetles called dermestids that really could be called museum bugs because their idea of a fine restaurant is a drawer of dried plants or animal skins or even other insects.

All the time the botany curator is identifying plants for people and saving plants for the collection, he is doing his most important job. That is research.

Your curator of botany should be looking for new facts all the time. Maybe he will try to find out how cactus stores water or how to make better-tasting corn or how to get more food from seaweed. Maybe he will try to find out all about the kinds of moss that grow in a deep gorge or on a mountaintop. Maybe he will study plants used for medicine. But when he does discover something no one ever knew before, even one tiny little fact, he writes about it for a scientific magazine that other botanists read.

His one little fact, all by itself, may not seem like much, but it might be just the one fact that is needed to put together something that has been puzzling another scientist.

6

Six-legged Science

One morning a woman went to a large city museum. She was elegantly dressed, wearing white gloves and carrying, ever so carefully, a small white box.

"Where will I find the gentleman who knows about insects?" she asked the guard at the door.

"Oh, you mean the curator of entomology," the guard said.

The woman waved one hand like a butterfly and said, "Well, whatever he is . . ."

The guard took her to see the curator of entomology, the curator who studies insects.

"Here, my good man," said the woman impatiently. "I found this creature in my apartment. It may be quite rare. Can you tell me what it is and if it is valuable?" And she gave the small white box to the curator.

The curator opened the box and tried very hard not to laugh. "Madam," he said, "this is *Periplaneta americana*."

"Oh my, how interesting," said the woman, looking pleased.

"Yes, most interesting," said the curator. "You have a cockroach."

"But, you said . . ." the woman blustered.

"I said the scientific name for cockroach."

The woman's smile disappeared. She glared at the curator and snatched the white box from him. "Well, of all the incompetent idiots! A cockroach indeed. Where would I get a cockroach? I shall report you to the director." And she left the museum never to be seen there again.

It is not always easy to tell people the truth. People do not want to know that cockroaches or fleas or termites can live in any house. Insects do not ask at the door for permission to come in. They can live in the house of people rich or poor.

Entomologists spend much of their time helping people identify the insects they have seen. They tell people how to get rid of carpenter ants that are eating the wood of kitchen cupboards, and fleas that decide to bite people instead of dogs. They know how to deal with wasps who have made their nests in the front entrance of a school or lice that have settled in long hair. Curators of entomology tell children how to keep cocoons, and they answer questions about fly sprays. They tell which poisons are safe to use and which ones would be dangerous to humans, birds, and other animals.

There are more than 900,000 different species of insects known, and some entomologists think that there may be almost that many we haven't yet found. So it is unlikely that any one entomologist can know about all the insects. Usually scientists decide to specialize in learning about one kind, such as beetles, but even that would be an enormous job. There are about 250,000 kinds of beetles.

Insects are very important in our lives. Without them we

would not have honey. Songbirds would have no food, and dead things would not decay as rapidly. Few crops would grow because the plants would not be pollinated. Insects even help control other insects. The cockroach, as miserable as he may be, does kill bedbugs. Insects are annoying, but they are necessary.

So, in order to learn as much about the insects as possible, your curator would most likely start a collection. But unlike some of the other curators, he might have two kinds of collections, one dead and one alive.

Many museums have live insect collections. This is a working hive of bees that can fly in and out through a window in order to collect nectar from flowers.

From a live collection of gypsy moths, for example, your curator could learn when and how they lay their eggs and how long it takes the eggs to hatch. He could find out what kinds of food the moths eat and how long they live. In other words, he could study the insect's life cycle, from birth to death. Such information might then be useful to a botanist studying the trees gypsy moths destroy as they feed upon the leaves.

From living collections entomologists can learn whether insects are attracted to colors or noises or odors. They can study how insects communicate with each other, the "language" they use. How does an ant tell other ants he has found a delicious piece of food too big for him to carry alone? How does a bee let other bees know where a good supply of pollen waits? Insect communities are organized so that each individual knows exactly what he must do. The only way *we* can find out what they do is by watching them.

But most of your curator's insects will be put into the dead collection for a different kind of study. Collecting insects requires simple, inexpensive, and often homemade equipment. For butterflies and moths, a lightweight, long-handled net is used. A heavier, shorter-handled net must be used to sweep through tall grass to collect grasshoppers and insects that live closer to the ground. Beetles and bugs are easy to trap in an empty soup can buried in the dirt with the top of the can level to the top of the ground. Grubs and worms as well as beetles stumble into the can trap.

Collectors know how to make insects come to them with lures. Some butterflies are attracted to decoys, imitation paper butterflies hung from an almost invisible thread from a tree branch.

Night-flying moths can be lured by "sugaring." A mixture

of mushy bananas and over-ripe peaches mixed with a bit of sugar is smeared on tree trunks. Peanut butter is a good smeary lure, too. After dark, the collector goes to the smeared tree with a flashlight and swoops the moths into his net. Because insects are attracted to light, it is easy to lure them onto a white sheet hung in front of a lantern at night. The moths land on the sheet and the entomologist can catch them easily.

The caught insects are put into a killing jar, which is a wide-mouthed bottle. On the bottom of the jar is a layer of cotton or sawdust soaked in cyanide or carbon tetrachloride, which kills the insects quickly.

The most important piece of equipment your curator will take on collecting trips with him will be his notebook. Even if he isn't exactly sure of the insects' names, the curator will write down all kinds of information. He will note the date, the time, the place, the temperature, the weather. He'll need to know if the day was rainy, sunny, cloudy, or foggy. He'll make a note of the ground conditions. Was the soil sandy, gravel, rock? Was it wet, dry, or in between? Were the insects in the grass? What kinds of plants were nearby? Were the insects near or in the water, and if so, was the water salt or fresh, stagnant, or swift-running? How were the insects caught . . . by hand, light, sweeping, dredging, trap? And then he puts his name, as collector, and any special remarks that might tell how the insects lived.

Back at the museum, the curator of entomology takes the insects from the killing jar and carefully spreads them for drying on a setting board. The board is made of pine, which is soft enough to push pins into, with a narrow slit down the center where the insect's body rests. The wings are spread on either side of the slit and pinned until they dry.

The dried insects are mounted in drawers and labeled. All of the insects are sorted according to families. All the beetles are together, all the praying mantises, all the ladybugs. Each species has its own place in a collection.

Of all the research going on with insects, probably the most important is finding out how insects can be used to control themselves. This is called biological control, and it started when we discovered that insect sprays were killing more than just insects.

When DDT was invented, everyone thought it was great and that it would keep insects from ruining crops. For a while it did. But many insects lived long enough to lay eggs that produced stronger insects than before. These stronger insects did not die when sprayed with the DDT. And then scientists found that the DDT itself was dangerous. It not only killed the insects, but it also killed or harmed other living things, including man.

When a cow eats grass that has been sprayed with DDT, some of the DDT stays in the cow's body. When we eat apples or peaches that were sprayed with DDT, some of the DDT stays in the muscles of our bodies. And it seems that the larger the animal, the more muscle it has and the longer DDT can stay in the animal's body doing damage, making him sick.

So entomologists, who had been learning how female insects attract males, had an idea. They found that some insects are attracted by a special odor, others by a sound. They learned to imitate the sound electronically and to make the odor from chemicals, and they attracted the male insects into traps. Now, if they just killed the males, other males would still mate and insect eggs would still be laid. Instead, the trapped males were treated so they couldn't help the females have babies. They couldn't fertilize the eggs. This is called making them sterile. Then the males were let go. They found females and mated as they always did. Only this time, because the males couldn't fertilize the eggs, the eggs didn't hatch. Thousands of insects were prevented from being born, and nothing else was hurt.

The world is full of people, and we need to produce food to feed everyone. The world is full of insects, too, and the insects feed upon many of the crops. Finding how to control the insects without hurting either crops or people is an enormous job. Entomologists can only do that job by learning everything they can about how insects live, how they help us and harm us.

7

On Top of the Museum

One of your curators may seem to be doing things backward. He may arrive at work when the night watchman does and leave at sunrise. You would seldom find him cleaning, sorting, labeling, or repairing a collection. His collection, for the most part, would take care of itself, and he would know exactly where to find it when he wanted to study it.

This person would be your curator of astronomy, and the objects of his collection are the sun, moon, stars, comets, meteors, all the things in our galaxy and beyond into the universe.

The men and women who are astronomers think big. While we think of how many blocks to walk to school or how many miles to drive to the country, astronomers think in light-years. A light-year is the distance light travels in one of our years in a vacuum, and it is about 5,880,000,000,000 miles. Distances to the stars are measured in light-years.

Your astronomer would probably have some of his big thoughts interrupted by strange calls. A common one is from a person asking, "Can you help me? I'm going on a trip, and I'd like to know if this is a good time in my horoscope for travel."

"I'm sorry," your curator would answer. "I think you want to speak with an astrologer. I'm an astronomer."

"Well, what's the difference? What *can* you do?"

"I'm a scientist. I can teach you about the stars and all the things in our universe. I'm a sky watcher. An astrologer is a people watcher. An astrologer tells people how the moon and stars can affect their lives at different times. They try to help people in making plans for their lives."

And people will call your museum almost every day to ask the astronomer, "Will the stars be out tonight?"

The stars are out all the time. We don't see them during the day because the sunlight is brighter than the starlight. We may not see them at night because there are too many clouds or the lights from city buildings and streets are brighter than the starlight.

Can you remember being in the country on a clear summer night? Remember looking up and seeing what seemed like trillions of stars, so many that it was hard to find the constellations? But when you returned to the city, the stars seemed to have disappeared. The telescopes in the observatories on top of city museums are useful only on clear nights. For that reason, many museums build their observatories far out in the country where there are no lights to dim the view.

The most useful kind of telescope in the city is the solar telescope. You can use it to look at the sun during the day. There are storms on the sun called sunspots. Astronomers have been watching and photographing these sun storms for many years. They have found that changes in the sunspots seem to go along with changes on the earth. Growth of trees and cycles in animals' lives seem to be affected by changes on the sun. Many more years of collecting facts at observatories will answer some of the questions about our sun.

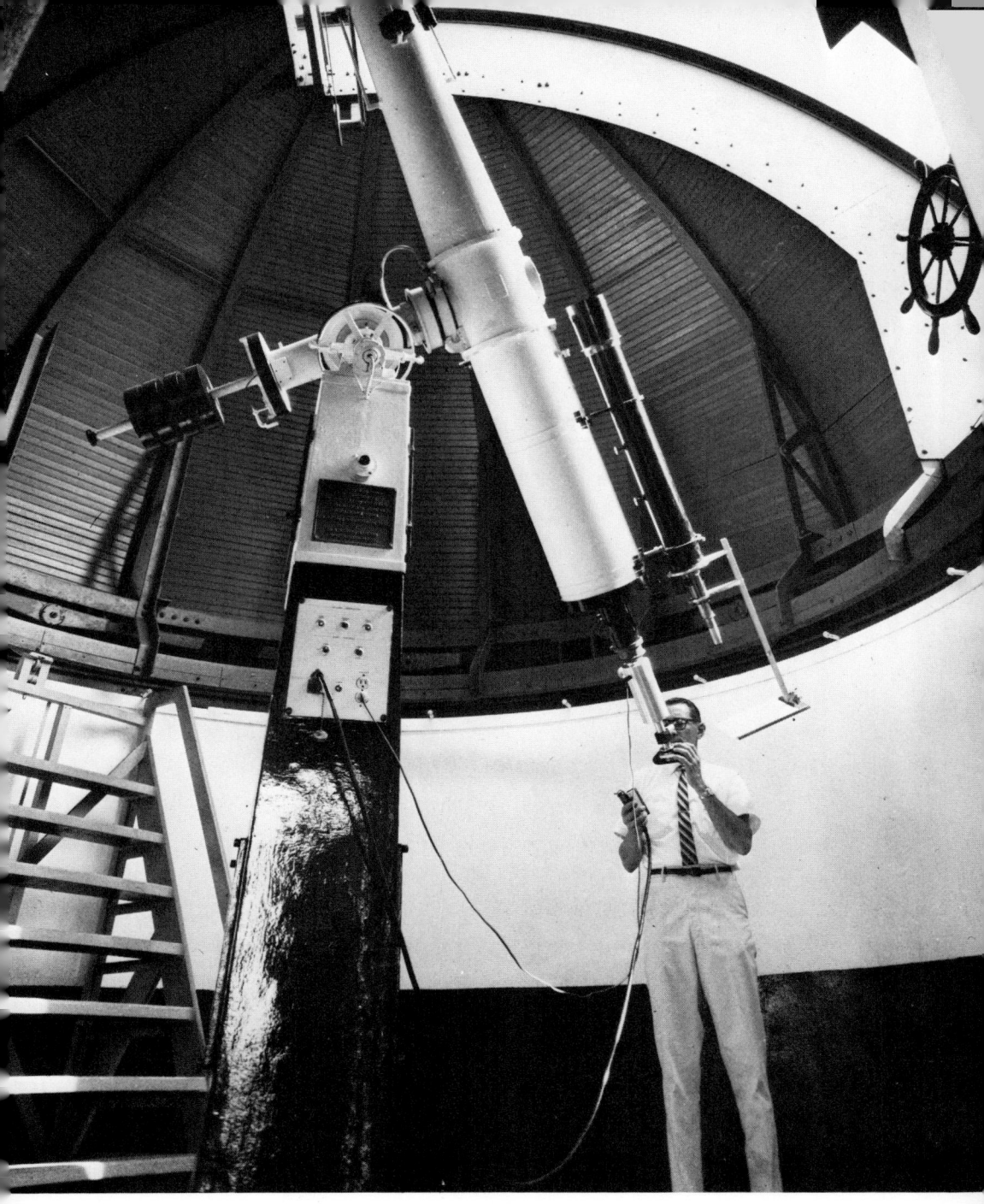

A telescope on top of a museum is one of the most important tools that an astronomer has to study the stars and planets.

Astronomy is one of the world's oldest sciences, but in many ways it is also the science of the future. For centuries, simple telescopes have been used to watch comets, meteors, planets, and phases of the moon. But for looking farther into space, astronomers needed more powerful telescopes. They built radio telescopes. You don't *look* through a radio telescope. You receive signals from outer space that tell about distances and objects that we can hardly imagine. Signals come back from other suns, from things called black holes, and from things we have no names for yet.

Your curator would be involved in teaching people about the universe. On clear nights, he would most likely invite the public to the observatory where they could learn about the moon and stars. But there would be so many overcast nights that your curator might suggest building a pretend observatory called a planetarium. A planetarium is just a room with a movie screen shaped like a bowl hanging upside down from the ceiling. In the center of the room is a round projector full of pinprick holes. When light shines through the tiny holes of the projector onto the overhead screen, it looks like stars on a black sky.

When you take your seat in a planetarium and the astronomer dims the lights, it seems as though the sun is setting. Your eyes get used to the darkness, and you begin to see the outline of the Big Dipper or the mighty hunter Orion. Even the Milky Way will appear like a band of stars across the sky. The astronomer can teach you how to navigate a ship by learning the stars in the planetarium or to recognize the phases of the moon as it makes its monthly tour across the sky.

To find out how a planetarium works, get a shoebox with a cover. Poke tiny holes in the cover with a needle. Inside the box put a flashlight. Go into a dark room and turn on the flashlight. Put the cover back on the box. You'll see stars on the ceiling in the pattern of the holes you poked in the cover. A planetarium can be as small as a shoebox or it can be an enormous, expensive theater with seats for hundreds of people. No matter what its size, it's only a pretend sky.

Astronomers collect photographs of anything they can see with a telescope. But they have only one collection of real objects. They collect meteorites, which are rocks from outer space. In fact, before our space flights and walks on the moon, meteorites were the only things on earth that came from outside our blanket of atmosphere.

You may have seen meteors. They are also called shooting stars, and they flash for an instant as they enter the atmosphere and burn brightly. When they land on the earth, they are called meteorites. Meteorites are bits broken off other planets or moons or comets. Some of them are only specks of dust and others are the size of pebbles and small rocks. Once in a great while a large one makes it through the blanket of air that protects us. In 1908, the Great Siberian meteorite exploded and flattened every tree for thirty miles around.

Centuries ago, when astronomy was a new science, it was believed that our sun was the center of the universe, the center of everything that existed. But astronomers have expanded that science so much that now we know there are trillions of stars that could be enough like our sun, with planets enough like our earth, to support life as we know it. We are not alone. Perhaps right now, way out in space many light-years away, farther than we can imagine, there are other curators of astronomy at museums trying to find life on another planet.

8

Skeletons and Skins

You might have a hard time finding your curator of zoology every time you need her. One morning you might see her coming in late, tired, wet, her muddy boots leaving a trail behind her.

"Where on earth have you been?" you might ask.

"Banding sea gulls," she'd tell you. "Went out at dawn. The sunrise was beautiful. Banded quite a few."

And she would explain that banding means putting a tiny bracelet around one leg of the bird. The bracelet has a number on it from the U.S. Fish and Wildlife Service. When the zoologist bands a bird, she sends that organization a report telling where and when she caught the bird and let it go again. Someday when another bird watcher sees that bird or finds it dead, he will send the bracelet to the Fish and Wildlife Service. The people there run the number through their computer, and it tells them where the bird came from. That way zoologists can keep track of how far birds travel and where they go at different times of year.

Then you might ask your curator of zoology, "What are you going to do with the dead sea gull I see sticking out of your knapsack?"

"Oh, I picked him up along the shore. I'll make a study skin of him."

Making a study skin is a way of preparing the skin so it can be studied. It can be measured, and the fur or feathers can be compared with others for color or shape. A museum has drawers full of study skins.

By examining rows and rows of bird study skins from around the world, a curator can learn how birds grow and change and migrate.

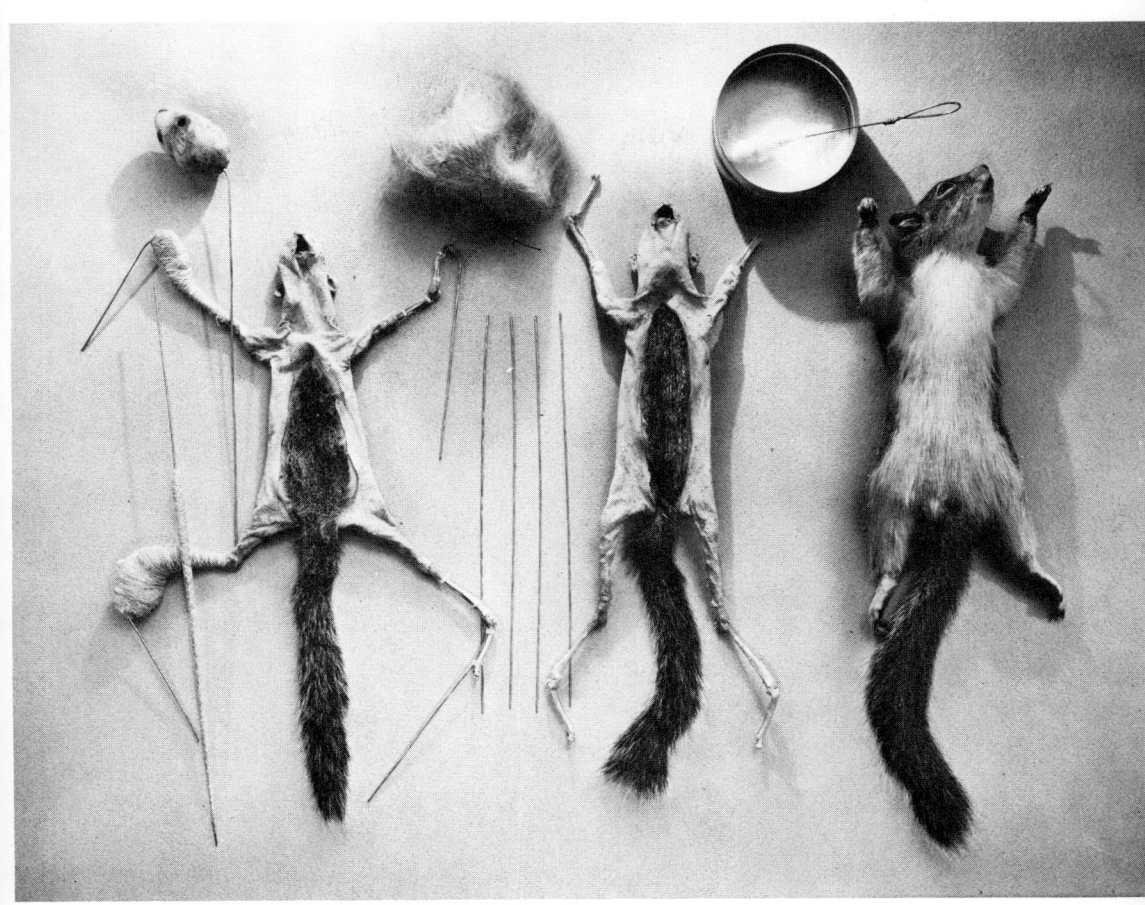

These are some of the steps in making a study skin.

When the curator makes a study skin, she usually saves the skull, too. When you take bones out of a dead animal during skinning, they are all sticky and still have some flesh on them of course. There are different ways to clean them. When there is no rush, a curator will let the bugs do it. The museum bugs, the same ones who like to dine on collections, can be quite useful.

After the animal is skinned and most of the flesh is taken off, the bones are put on shelves or in a box filled with black beetles about the size of your little fingernail. These beetles lay eggs that hatch into inch-long white worms called mealworms. Maybe you've fed mealworms to a pet lizard.

The mealworms eat the flesh off the bones. When most of the bone is clean, the curator takes it out of the bug box and bleaches it in Clorox or in the sun. She tags the skull with the name and date and location the animal was found in.

Whole skeletons are put together from bones cleaned like this. If she wants a faster job, the curator can cook the bones or put them in strong chemicals.

While your curator of zoology may be banding birds and studying sea gulls, other curators of zoology may be crawling through caves to learn about bats, or diving in the ocean to study sharks. One zoologist may be freezing in the Arctic as he tags polar bears to see how far they roam, and another might be roasting in a desert studying rattlesnakes. And some may never even leave the lab. They may be studying the collections they have.

When the entomologists were worrying about DDT and how it affected insects, the zoologists were worrying about how DDT hurt larger animals.

They had the idea that somehow DDT hurt female birds so that, when they laid their eggs, the eggs were not sturdy. The shells were very brittle and broke as soon as the mother bird tried to sit on the nest. How could the zoologists prove that it was caused by the DDT? One way was to study eggs that had been laid before DDT was invented.

Who keeps old eggs? Museums, of course. From collections of empty eggs in museums, the curators would weigh and measure the thickness of eggshells. All the eggs laid before birds ate DDT-sprayed insects and plants were stronger.

That piece of research was important because it helped stop DDT. Zoologists are always working to help animals survive, to keep them from becoming extinct.

Remember Little Red Riding Hood? Everyone who heard that story learned that wolves are fierce beasts that eat people. It's hard to forget things you learn when you're small. All the people who grew up remembering the awful wolves thought it was fine to kill wolves. In many places the governments even paid bounties for wolves. A bounty is money paid to a person who brings in the skin of the animal to prove he killed it.

Not long ago museums exhibited wolves with lips curled back, teeth bared, as though they were ready to attack. But some zoologists began to notice changes in areas where many wolves were killed by bounty hunters. With fewer wolves around, there were many more mice and rabbits and lemmings. And botanists reported that, at the same time, many young trees were destroyed. What was the connection?

As the zoologists watched the wolves, they learned that the old stories were not true. Wolves did not kill many deer and moose. Mostly, they lived on a daily diet of small animals, especially mice, rabbits, lemmings, and other rodents. When there were not as many wolves to eat the small animals, the families of mice and other rodents grew and grew. Small rodents can have a batch of babies almost once a month. And they all need food to survive. Suddenly, many more little animals were surviving because there were fewer wolves around to kill them. There was not enough food for all the small rodents. And so they began to gnaw the bark of young trees. They chewed the bark all around the tree, causing it to die. What happened was a kind of chain reaction. It is what zoologists mean when they talk about ecology, because ecology is the study of how all living things are related. In this case, the chain reaction involved wolves and rodents and trees.

Wolves can kill large animals, but they don't do it often. A pack of wolves, a family traveling together, will follow a herd of caribou. The wolves can't run as fast as the caribou, so they never catch the strong, healthy animals. But when the herd of caribou is running, the old or sick caribou fall behind. Then the wolves can catch one of those. The Eskimos say that the wolves keep the caribou herds healthy. What they mean is that by killing off the old and sick caribou, the wolves leave the healthy, strong ones to breed and produce young healthy animals.

Some of the nice things about being a curator in a museum are the surprises. You never know where information may come from. Sometimes even youngsters help. One day a group of children were hiking through Shale Creek, a nature sanctuary near Buffalo, New York. As they poked around the creek bed, they found crayfish and newts. Because the children were members of a museum hiking club, they knew the right way to poke around. They knew that when they turned over a rock, they had to put the rock back in place. It might be the home for a newt or crawly thing. They also knew that a nature sanctuary is a place where you take nothing with you and leave nothing behind.

Suddenly one of the boys hollered, "Hey, I found an orange frog."

"Don't be silly, frogs aren't orange," said one of the leaders.

But the frog was as orange as an orange.

The teacher said, "For heaven's sake. That's *Rana clamitans*. That's a green frog."

"No, it's orange," the boy insisted.

"Yes, it's orange, but it's green," said the teacher. "What I mean is, this is a common green frog, but something happened when it developed so that it looks orange. Its scientific name is *Rana clamitans*." And she showed the children the frog's picture in the field guidebook.

They broke the rule of nature sanctuaries then because they thought it was important for the museum zoologist to see this strange amphibian. At the museum, the curator of zoology told the children that every animal, including man, has coloring in the skin or fur or hair. Coloring is called pigment.

Sometimes animals are born with no pigment at all. They are all white, but their eyes look pink because the blood vessels show through. An animal with no coloring is called *albino*.

The green frog simply had too much yellow pigment. The zoologist gave the frog's unusual color the long name xanthochroism, which means yellow (xantho) color (chroism).

The orange frog lived at the museum for a while, well fed on juicy flies. When it died, a wax model was made that looked exactly like the live frog. From that small exhibit people learned facts they had never known before. Exhibits do not have to be large or expensive in order to be interesting, and animals do not have to be in faraway countries in order to be interesting to a zoologist.

9

Skinning the Cat

As a good museum director you would look around a lot. You'd visit the curators and find out what was going on in every part of your museum. You might see the curator of zoology struggling with the dead llama that had been sent over from the zoo.

"Everytime I see you," you'd tell her, "you're making another study skin. Why don't you make some of the animals look alive?"

"I don't know how," she'd tell you. "I can do the study skins, but I'm not an artist."

"What kind of artist makes animals?"

"A taxidermist," she'd say.

Taxidermy is a strange word. The first part comes from the Greek word *taxis* (not pronounced like taxicab, but like tacks -iss) that means arrangement or preparation. The second half is from the Greek word *derma*, which means skin. So taxidermy is the preparation of skin.

The first animals fixed for display were really stuffed, just like teddy bears. The taxidermist would hang the animal head down and ram straw into the body. The animal looked rather lumpy and not much like it did while alive.

At one time, taxidermists built wooden frames and then covered them with animal skins.

From that the taxidermist moved on to building animals out of wooden slats over wooden ribs, the way a canoe is built. The animal's skin was stretched over the wooden frame. The animals began to look a little better, but still a bit stiff and wooden.

Each taxidermist tried a different way to make animals look real, and in the 1900s it became very popular to have a stuffed animal of some kind in your living room. Ladies put stuffed pheasants on their pianos, and gentlemen hung deer heads in their dens.

There were some taxidermists who thought it was nonsense to try to make an animal look alive. One even wrote a paper telling how it would never be possible.

A man named William Hornaday worked as a taxidermist in Rochester, New York, for a company called Ward's Natural

Science Establishment. He put together the first really lifelike group of orangutans to show at a meeting of scientists in 1879. Everyone at the meeting liked them, and a man who was helping put together the American Museum of Natural History then bought the orangutans for the new museum.

Those were exciting years for museums. People like Mr. Hornaday and Frederic Lucas were making exhibits. Mr. Lucas worked at Ward's, too. He later became the director of the American Museum of Natural History, but he started at Ward's when he was a boy. His father had written to Henry Ward asking what he would suggest doing with a boy who seemed to be good at nothing except skinning snakes. Mr. Ward wrote back, saying, "Send him on."

But of all the people who did great things for taxidermy, the most famous was Carl Akeley. When he was growing up, he sent for a mail-order course in taxidermy and taught himself how to mount animals. Even when he got a job at Ward's, he spent all his spare time at circuses, zoos, and county fairs, watching animals as they moved. He took pictures and made sketches, teaching himself how animals really looked.

One day there was a terrible accident that gave Carl Akeley a chance to show how really good he was. The P. T. Barnum Circus was traveling in Canada with their famous giant elephant Jumbo. At that time, Jumbo was the largest elephant ever captured. After the last show the elephants helped load the trains by pushing wagons up on them.

Jumbo was being led across the railroad yard with Tom Thumb, a baby elephant. Suddenly a train no one expected came roaring down the track. The engineer saw Jumbo and tried to stop. Jumbo, trying to protect the baby elephant, lifted his trunk and trumpeted a warning. Then he charged the train.

mâché. Jumbo was delivered to the circus ready for the opening parade on March 4. He rode for many years at the head of the parade on a brightly colored circus wagon.

Jumbo's papier-mâché skull began a new kind of taxidermy. Taxidermists became artists. Where once they had made a stiff model of wood or wire and pulled the skin on, taking tucks and making wrinkles if the skin was too big, now they became sculptors.

Some taxidermists begin with the skeleton of the animal and build a papier-mâché animal over the bones. Others make an armature, a wire and wood skeleton like sculptors use.

Today taxidermists make animal models that are like sculptures. When covered with the real animal skins, the models look alive.

This fiberglass model looks real, even though it isn't.

They build muscles of papier-mâché until it looks like the animal without fur or feathers. The carefully tanned skin is put on over the model and sewn up so neatly that no one can see it.

So when you see an animal in a museum that looks real and you ask, "It it real?" you know the answer is "Partly." His bones are real, and his skin is real, but usually his tongue and eyes are made of plastic because they are soft and would rot if they were real.

Taxidermists work in places other than museums, too. Hunters and fishermen take their catches to taxidermists to be mounted.

Deep-sea fishermen like to have marlin and enormous tuna mounted to hang in their homes. What they don't know is that the fish they have hanging over the fireplace is not always the real fish but may be a fiberglass model painted to look real. Sometimes companies that mount fish have a stack of different size fiberglass fish. They choose one nearest the size of the fish brought to them and paint it. They put a sign under it telling who caught it and how much it weighed and give it to the happy fisherman.

Museums are using fiberglass, too. In the Smithsonian Institution in Washington, D.C., there is an enormous blue whale. It is made of fiberglass, and everyone is happy about it. A fiberglass whale is much lighter than a real one would be, and they can hang it from the ceiling. And by having a fiberglass whale, they don't have to kill a real one. You can still be thrilled at the size and strength of the great blue whale without a whale having to give its life for your education.

10

Frozen Snakes and Spiders

"Good morning," you might say to the taxidermist one day. "The wolves you mounted look great. What are you building this week?"

"Well, Chief, I want to talk to you about that. Remember when I took that trip and I visited those other museums? Well, I saw this great thing called a freeze-drying machine. Did you ever taste freeze-dried coffee or freeze-dried mushrooms?"

"Yes," you might tell him, and add, "When we went camping last summer, we took freeze-dried pork chops and even some freeze-dried ice cream. It looked like powder, but it tasted like real ice cream."

Freeze-drying is a way of keeping food from spoiling. It is a process to take the moisture out of food. Freeze-dried food takes up less space than canned goods, and it doesn't need to be kept cold like frozen food. Some museum people saw all the freeze-dried food and thought: If you can freeze-dry mushrooms for eating, why not freeze plants and animals for a museum?

Museums get dead animals all the time. Some come from the zoo, and some are found dead along the roadside. The taxidermist has to skin and gut them right away so they don't rot. If he doesn't have time, he puts them in the freezer. When he needs the animal for an exhibit, he has to defrost it and skin it. If the museum had a freeze-drying machine, it would save time. The taxidermist wouldn't have to skin and gut the animal at all.

If a taxidermist found a dead cardinal, for example, he would wash it in soap suds and then, when it was dry, he would brush the feathers into place very gently until the bird looked fresh and new. He would then put wires along the cardinal's legs and into its body in order to perch it on a branch. Then the bird would be put in a freezer. When it was frozen as solid as an ice cube, the taxidermist would put it in the vacuum chamber of the freeze-drying machine.

Museums use freeze-drying machines like this one to prepare small animals for exhibit.

The vacuum chamber is a place where there is no air, just nothing. It is space. The vacuum pulls all the moisture out of the bird. Every day the bird would weigh a little less. The taxidermist would weigh it every few days. When the bird weighed about one third what it did when it was alive, he'd know it was ready. All the moisture would be removed. It would feel like a piece of foam plastic, but it would look alive.

The whole process would take a couple of weeks. And while the bird is freezing and drying, the taxidermist could be working on other animals.

You might wonder whether it's possible to freeze-dry an elephant. Right now there aren't any vacuum chambers big enough. The freeze-drying machine's best use is for small animals that are difficult to preserve other ways, things like toads, frogs, snakes, and spiders. Your taxidermist could make a rattlesnake look as though it were striking, although he might have to touch it up with some paint. Some of the natural colors of animals fade as they dry.

Other people besides the taxidermist could use the freeze-drying machine. The curator of zoology could make study skins in it, and doctors might be interested in the machine, too. If a doctor took out your tonsils, he could freeze-dry them. Then when he wanted to teach a class of doctors about tonsils, he could show them what a diseased one looked like. Doctors could freeze-dry many parts of the body for teaching.

And so as director of your new and modern museum, you might agree with the taxidermist and get a freeze-drying machine.

11

Putting It All Together

Well now, would your museum be all together? You'd have curators caring for all the treasures. And you'd know, down to the last African trade bead, exactly what was in your museum. You would be so organized that you'd be almost afraid to put a paper clip down for fear someone would put a label on it.

Then you would invite your friends to visit. And your best friend would walk up and down the rows of cabinets with closed drawers, past the glass cupboards full of birds lined up as though they were waiting for a bus, and he would say, "How dull."

"What do you mean, dull?" you'd ask, more than a little annoyed.

"Who wants to look at a hundred arrowheads lying in a row? Who wants to stare at fifty-six fossils in a drawer?"

"But we've done everything a museum should do. We've saved things so people a hundred years from now will know what treasures have been on this earth. It's all labeled, and our curators are studying, looking for new facts."

"Who cares?" your friend would ask.

Sometimes looking at rows and rows of fossils or shells can be boring. That's why modern museums have different kinds of exhibits.

"All of my museum people care."

"That's not enough. I thought you said a museum was like a real encyclopedia, that people could learn just by being in a museum." And then he would walk over to a row of birds and point at a big, bald one and say, "What's this?"

"It says right on the label that it's a turkey buzzard." You'd point with pride.

"But what does it eat? Does it live in a cave or on a mountaintop?"

"Read the label," you'd almost be ready to scream at him.

"I hate labels," he'd tell you quite honestly. "And most people do. Why, I'll bet you don't read any sign smaller than a billboard."

Putting It All Together

By this time you would be angry, but you'd also be thinking. Much to your friend's surprise you'd say, "Thank you. You've given me an idea."

The next morning you'd call a meeting of all the people who work in your museum so far, the curators, the registrar, and the taxidermist. And you would tell them that although you have beautiful collections, the museum is dull. And dull is worse than ugly or cluttered.

"What can we do?" one of the curators would ask.

"I've been thinking about that," the taxidermist might say. "I heard some children talking about that muskrat standing on the shelf. They didn't know where a muskrat lives or how it finds food. Couldn't we show them?"

This exhibit of woodland animals shows not only what a skunk looks like, but also where it lives.

"That's it," you'd say. "A museum should be a big show-and-tell. We'll put the animals where they belong, the polar bears in the snow, the beavers in the pond, and the gorillas in the jungle. Instead of rows of arrowheads, we'll make villages with pretend people making the arrowheads and wearing the clothes of their tribe and cooking the kinds of food they ate. We'll let our visitors pretend they are really there, in the desert or the jungle."

But there are only a few people who work in your museum, and they don't know how to build exhibits, so the museum will have to hire artists and carpenters and people who make the pretend plants. You'll need designers and model builders. And as you begin to plan your exhibits, you will learn that building lifelike dioramas is like putting on a play.

Were you ever in a play at school? At first everyone was standing on the bare stage, holding papers to read from. They all wore their regular school clothes. It was hard to imagine that it would ever turn into anything you'd want your parents to come to. And then it gradually changed. First there were costumes, then special lights, and properties—all the furniture and things the actors use on the stage. At last it was a play, a bit of pretend that seemed very real for a while.

Building an exhibit is like producing a play or making a movie. It is telling a story. In a play about a princess, the princess doesn't have to wear a sign telling you she is a princess. You know it by her costume and by the elegant palace in which she lives. In a museum you know the people in an exhibit are cavemen or Eskimos because they are wearing costumes and they are in a cave or on a snowy ground.

When a museum is going to plan a new exhibit, they have a meeting of all the special people. Scientists, who know the

People with various skills work on museum exhibits, including this woman who is painting plastic leaves for trees.

facts, work with designers who have ideas, with artists who can paint and build models, with craftsmen who can build cabinets, with taxidermists who build animals, with writers who write labels, and with photographers who make murals.

You might decide to show a meadow at early evening so that city people could know what it was like in the country.

The curator of zoology would say, "How about a mother skunk and three babies trotting after her?"

"And a big, fat toad," the taxidermist would add.

"We'll need goldenrod and asters and different grasses," the botanist would say. "We'll have to hire preparators to make plastic or wax flowers and leaves."

The curator of entomology would suggest, "There should be a monarch butterfly lighting on some milkweed, and maybe some crickets and grasshoppers, too."

Scientists, artists, designers, and builders are studying a model of an exhibit they planned prior to starting the actual construction of it.

When the fake trees, the fake snow, and the wolf model are put together, this exhibit will look real.

"Crickets," you'd say. "If we're going to have crickets, why not let people hear them chirp? Nothing nicer than the song of the crickets at evening."

"How will we do that?" one of the people would ask.

"We'll go out in the field and tape-record some crickets, and maybe some bird songs, too. Then we can play the tape behind the exhibit."

As your meeting went on, you'd decide on pheasants and maybe a nest with eggs and all sorts of other things that go in a meadow. You'd make a list of the people you'd need to hire to help make your exhibit.

Everyone would leave the meeting anxious to get started on the job they had for their part in the exhibit. Times would be set for other meetings and for a final date when the exhibit would be ready.

At last, when everything was finished and all parts of the exhibit were brought together, they would be carefully assembled.

Some museums use sound effects and temperature changes to help people imagine they are part of the exhibit world. You might walk into a cold room where polar bears are sitting on an icy-looking ground. Some museums use animation, as Walt Disney does in the exhibits at Disneyland. In some places you can push a button and see the tentacles of a squid move as though he is looking for food.

Whatever the exhibit, it has to be right because exhibits are too expensive to take apart very often for changes. The story they tell must be worth telling for a long time.

Every museum wants its visitors to say, "Oh, wow!" or "I didn't know that!" A museum wants people to come back often to look and learn.

12

Gorillas Are Great

The African Hall at the American Museum of Natural History in New York City is one of the most exciting exhibits in any museum. In one enormous room you can wander from the edge of a water hole to the waterless desert or into the darkest jungle. Building that exhibit was a super job of putting together the talents of taxidermists, artists, collectors, and all the other museum specialists. The idea came partly from a gorilla.

Gorillas are great; both kinds of great, great-huge and great-wonderful. They are the gentle giants of the rain forests. But people didn't always know that.

Gorillas used to be called the "wild men of the woods." It started when a French hunter and explorer, in 1855, became the first white man to shoot a gorilla. Like a fish story, his adventures grew larger and larger the more he told about them.

Nobody wants to be known as a hunter who shot a helpless animal. It is much more exciting to have people think the animal was fierce and dangerous. And so the hunter told people about the man-eating, screaming monsters that came roaring out of the jungle to attack.

Not long after that, an artist made a bronze life-size statue of a fierce-looking gorilla carrying a woman in his arms. That statue stood in the basement of the American Museum of Natural History near the elevators. Every time Carl Akeley, the taxidermist, waited for an elevator, he saw that statue. He got angrier and angrier.

"It's a lie," he told people. "Gorillas aren't like that."

Now, Carl Akeley had never seen a gorilla in the wild. There were few movies about gorillas, and none with live gorillas in them. There were no television programs showing animals as they really are. But Mr. Akeley knew a lot about animals, and he believed that an animal will not attack man unless it has to protect itself. He knew, too, that an animal might fight fiercely when it is cornered or threatened, but that whenever it can, an animal will run away. Mr. Akeley was determined to get movies of real gorillas and to build an exhibit that would show a gorilla family as they live in the African mountains.

In 1921 he made his first trip to the slopes of Mount Mikeno in Africa, where gorillas lived. Expeditions in those days went on foot, with native people carrying the supplies, including the tents, chairs, cooking equipment, and everything the group would use. It was not an easy trip. On the second day out, they were climbing a slope that went almost straight up, like a wall. They saw the bushes move and knew gorillas were feeding along the ridge.

Carl Akeley cut his way through the thick jungle growth to get near the gorillas when suddenly there was a roar that gave everyone goosebumps. Carl Akeley was moving slowly along the ridge of the deep ravine with only a small four-inch tree between him and the rocks in the bottom of the chasm.

Gorillas Are Great

The second roar was closer, but they still couldn't see the animal making the noise. Later, when Akeley knew the ways of gorillas better, he found that the roaring was the gorilla's way of telling them to get away, to leave him alone. Finally, when the gorilla saw that no one was leaving, he roared again and charged out of the thick trees like a tank.

Carl Akeley raised his gun and shot. A huge silver-gray gorilla hurtled at him, stopped only by the small tree at the edge of the ravine. That tree kept the gorilla from crashing into the rocks. Mr. Akeley was happy that he didn't lose the 400-pound animal, that it would become part of the African exhibit and in that way would be useful.

Carl Akeley shot only animals he needed for exhibits. This giraffe was for his African Hall.

Getting that animal back to their camp was a terrible job. They spent the next day skinning the animal and making plaster casts of the hands and face so they would be able to make the gorilla as real-looking as possible. Because he wanted to show a gorilla family, Carl Akeley went back to shoot a few more animals. One of them did fall into the canyon, and even though no one else in the group wanted to take such a dangerous climb down the steep canyon walls, Akeley went after the animal. He had to stand on a narrow, slippery ledge to skin the animal. The native people helped him carry the heavy skin and bones up the rocky walls and back to camp. When others in the group said, "Leave it there," Carl Akeley told them that he didn't want to waste such a magnificent animal. "You only shoot an animal for food or study," he told them.

After they had the gorillas they needed for the exhibit, Akeley took movies, the first ones to show gorillas as they live. He took pictures of a mother gorilla lying on a branch eating leaves, with her baby playing nearby.

Later, an artist from the museum was sent to Mount Mikeno, to the spot where the gorillas had been taken, to copy the trees, the hills, the same blue sky so that he could paint the background of the museum exhibit to look like the great apes' real home.

While Carl Akeley was collecting the gorillas, he found that hunting gorillas was becoming a favorite sport for men rich enough to travel to Africa. A prince had killed eighteen gorillas on one trip. That made Akeley very angry. He worked hard talking to people in government until he convinced them to set aside part of the African jungle as a national park where no gorillas could be shot.

Gorillas Are Great

When Carl Akeley finished building this gorilla, he put it in an exhibit that contained replicas of the mountains of Africa.

Carl Akeley died before the gorilla exhibit was finished at the museum. He was buried at his favorite place, high in the African mountains where the gorillas roam.

13

Dinosaurs

One day you might look around your great new museum and say, "Something's missing. We don't have a dinosaur. A natural history museum without a dinosaur is like peanut butter without jelly."

No one had seen a dinosaur skeleton or even heard the word dinosaur until Mary Anning found the first dinosaur skeleton in England in 1811. Mary was twelve the summer she found it. She lived near the ocean shore, where she and her father collected shells in the summer. They were looking for shells when she found a huge, whole skeleton just lying there in the rocks. No one knew exactly what it was. Scientists who looked at it decided to call it Ichthyosaurus, which is a combination of two Greek words that mean fish-lizard. That's what it looked like, part fish and part lizard.

A few years later, after more skeletons had been found, one scientist decided that all the lizard-like monsters should have a name. He called them dinosaurs, which means terrible lizards.

After you talked with the geologist at your museum, you would know that finding a dinosaur is partly luck and partly knowing where to look. There are fossils on every continent and in all kinds of places—deserts, mountains, ocean shores, and mines. The people you send out would have to know what kinds of rocks are most likely to have dinosaur bones in them. They know this by the age of the rocks and the manner in which they were made. You wouldn't be likely to find a skeleton in a layer of rock made by a volcano pouring out its hot lava. But you would find bones in layers of rock that had been sand and clay millions of years ago, places where dinosaurs had lived.

The people who dig for dinosaurs are called paleontologists. The first part of the word, *paleo*, comes from a Greek word that means ancient. It is the study of ancient life, the study of fossils. Paleontology is a branch of geology because you need to know about the formation of the earth in order to study fossils. It can also be a part of zoology, called paleozoology, the study of prehistoric animals.

If you talked to people who dig for dinosaurs, you'd hear words like "hot, long, slow, tiresome." Digging is all of those things. It means using a pick and shovel to dig into the ground and uncover rocks and bones. Then it means being careful not to hurt the bones with the pick, so you switch to smaller tools such as trowels and whisk brooms. You work like the archeologists.

Can you imagine being the first person on earth to see an animal that was alive 400 million years ago, long long before people lived?

Roy Chapman Andrews, the man who led the expedition to the Gobi Desert in 1922, when they found the first dinosaur eggs, told people that "There is a thrill like nothing else on earth in discovering something new. I had that thrill out

Roy Chapman Andrews led the expedition in which the first dinosaur eggs ever seen by human beings were discovered.

there in the desert when I realized that lying before us were the first dinosaur eggs ever seen by human eyes."

And the really exciting thing about the dinosaur eggs was the solving of a question. Until that time paleontologists had only pieced together a few facts that told them dinosaurs laid eggs. They knew that dinosaurs were reptiles, similar to those alive today, and there was a good chance that the smaller dinosaurs, at least, would have laid eggs. Roy Chapman Andrews and his group knew the eggs they found were those of a dinosaur because they were found in layers of rocks holding dinosaur remains and no other kinds of animals. The shape and texture of the eggs told them they were reptile eggs.

If you were on an expedition with your museum paleontologist, and perhaps the geologist and the zoologist, too, you would have a team of experts to find dinosaurs. After you uncovered your dinosaur bones, you'd have to protect them right away. For millions of years they were protected by the rock they rested in and the blanket of soil on top of them. When suddenly exposed to the air, they may fall apart. You would work as the archeologists do to protect the artifacts they find. You'd shellac the bones. Then you'd dip strips of burlap in wet plaster and lay that over the bones. You'd make a plaster cast for each bone, the way a doctor makes a cast for a broken arm. The plaster casts, with bones snugly inside, would then be packed in straw and put into wooden crates for the trip to the museum.

Back at the museum, everything would have to be undone. You and the paleontologists and assistants would carefully chip off the plaster casts. Each bone would be dug or drilled away from the matrix. Matrix is the name given to the rock around the bones. You would use dentist's drills and tiny chisels for your work.

As each bone is cleaned and separated from the matrix, you would tag it with its name. Sometimes dinosaur bones are mixed up. If two dinosaurs died fighting each other, their bones would be in one pile. Or bones might wash down a river to be buried in a heap with lots of other bones. Before you put a skeleton together, you have to know which bones are which.

Next, you'd draw an enormous picture on papers on the floor so you could lay the bones out, like fitting a puzzle together. Where bones were missing, you'd have to make new ones of plaster or plastic.

Finally, you'd assemble the dinosaur the way you would put together a plastic model. The job is longer and harder

When dinosaur bones arrive at a museum, they must be cleaned and separated from the rock around them. Broken pieces are glued together.

This Allosaurus was strung together before it was finally mounted so that the curators could decide in which position it would look best.

than a small model, but it's fun because you're putting together something few people have ever seen, something from another time.

Building a dinosaur is slow but not too difficult. There is one huge problem, though. Building a dinosaur costs lots of money—more money than your museum might have. It can cost hundreds of thousands of dollars to send a group of people into the field for months to find, dig, pack, ship, and reassemble a dinosaur.

"Good grief," you might moan. "We don't have that much money. But we really want a dinosaur to show people. What can we do?"

You can buy a fake dinosaur.

"What?" you might scream, because you run a museum where things are real. "Have a fake? Not us!"

But there are good reasons to have a fake. There's the money, of course, or rather the not having money. And there's also the fact that there aren't many dinosaurs around where you can get to them. Most countries are protecting their own treasures. They are not letting teams of collectors from other countries move in and take out such valuable things as mummies or dinosaurs or buried treasures. If a dinosaur is discovered in Canada, a Canadian museum should have it. Treasures found in Mexico should be in Mexican museums. In the past, American and British explorers went to China or Africa to find things for their museums, but they cannot do that anymore. So it is getting harder and harder to find dinosaurs.

There is a third reason to use fake dinosaurs. That is to protect the real ones. If you had a real Stegosaurus, you wouldn't want it damaged. You wouldn't want people to break off parts of it or write on it with a ball-point pen or carve it with a knife.

And even if you protected it from people, you'd still have to worry about dust and chemicals in the air that eat away at bone and rock little by little.

At the American Museum of Natural History in New York City, they have a real Stegosaurus. There are only five of them in the United States. Stegosaurus is about 150 million years old. He's the one with the pointed shields along his back and down to his tiny head. The Stegosaurus was collected

This is the Stegosaurus skeleton put together at the American Museum of Natural History and copied for other museums.

in 1901 at Medicine Bow, Wyoming, and it has been on exhibit at the museum for more than forty years.

One day some visitors from the natural history museum in Osaka, Japan, asked if they could have a copy of the Stegosaurus. It seemed like a new adventure for the museum people, and they said, "Sure, we'll try."

The first thing they did was take apart the real Stegosaurus skeleton very carefully. There were more than 100 pieces. They built the model right in the dinosaur hall where visitors could watch. They didn't take it to one of the workrooms. For a whole year, the making of the skeleton was an exciting exhibit. Visitors even helped. Some sculptors and other artists volunteered and so did some high school students.

They had to paint each bone with latex, which is a liquid that turns to rubber when it dries. They put fifteen coats of latex all over each bone. The dry latex, or rubber, peels off the bones, forming a mold of each bone.

They poured liquid fiberglass into each mold. That made a fiberglass bone to match each real bone. They put all the fiberglass bones together just like the real skeleton.

When they finished, the new Stegosaurus looked so much like the real one that it was hard to tell them apart. The bones looked and felt exactly alike.

The Japanese museum people were delighted. In fact, the American museum people were so happy with the fiberglass Stegosaurus that they made a fake one for their own exhibit hall. Then they put the pieces of the real Stegosaurus that people had looked at for forty years in storage, where they will be protected from dirt, mischief, and temperature changes that can crack old bones.

The farmer would be glad to see you, and he'd tell you about the bones. "I decided to plant this land that hadn't been used in forty years," he'd say. "A drainage ditch had to be cleared and deepened, so I hired a backhoe, you know, one of those digging machines. The backhoe operator was scooping out the muck when he spotted the long bones. He cleaned them off a bit and tossed them in his pickup truck. We thought maybe they were dinosaur bones. They're sure big. Could be elephant bones, though. Circuses have gone through here in the past. Could be they buried an elephant that died."

Near Albany, New York, this story really happened. The bones were those of a mastodon, the elephant's hairy cousin who lived in North America about 10,000 years ago. There must have been thousands of the big beasts in the small part of North America that became New York State because about a hundred skeletons have been found.

The bodies of most animals are destroyed soon after death by scavengers and by rotting. Fewer than one out of a hundred animals become fossils. Most vanish without a trace. It was lucky for today's fossil finders that some animals were buried quickly, preventing them from rotting fast.

The bones of a mastodon would have had to sink quickly into fine clay, called marl, or into peat bogs, where they would be sealed off from the bacteria that make them decay. Usually only teeth and bones of an animal are found because flesh decays first. Sometimes crushed twigs are found stuck in the teeth of a mastodon or in the body of the skeleton, telling us that he had eaten the twigs.

Before settlers came to America and found bones here, bones of giant animals had been found on the European continent. In Siberia a group of people called the Tartars called these huge animal remains "mammots," which meant giant in their language. The word mammot was used everytime a similar skeleton was uncovered, and it finally changed into the word we use, mammoth. So the mammoth was the first name given to an ancient animal. Later, when comparison was made of the elephant-like animals of Europe and those found in this new land, one man noticed a difference in the teeth. He thought they had small points on them that looked something like breasts, so he named them mastodons. The first part of the word, *masto*, means breast. From that time, the American mammoths with the bumpy teeth were called mastodons.

Thirty-nine woolly mammoths and mastodons have been found frozen in the far northern parts of Alaska and Russia. Four of them were so quickly frozen after they died that they still had all their hair and skin. Even the food was frozen in their stomachs and blood in their veins.

The first time anyone in America found bones from a mastodon and wrote about it was in 1705. An article in the *Boston News Letter* told readers that they "thot it was the tooth of a giant man drowned in the Flood."

The "flood" meant the forty days and forty nights of rain that people believe inundated the land in Biblical times, causing Noah to build his ark. Churches and schools taught that the earth was 4,004 years old and that anything such as bones or footprints in rocks had to be from animals God destroyed. They did not believe that those ancient animals had anything to do with life on the earth now or at that time. The men who were finding these bones were beginning to suspect that the dead animals were the ancestors of modern animals, but it wasn't easy for them to go against their church belief and say so. For many years bones were found, and the explanation was always the same. One farmer found the print of a three-toed animal's foot, and the minister he showed it to told him it must be the footprint of the raven on Noah's ark. It had flown out to look around and must have landed on the rock, leaving a print. Many years later, scientists realized that the print belonged to a dinosaur.

Almost a century later, in 1799, a farmer in the New York region found large bones in a swamp on his land. A hundred neighbors helped him dig them out, but as they worked, the hole kept filling with water. They were just about to quit when Charles Peale, an artist who had started his own museum in Philadelphia, said he'd finish the job if he could have the bones for his museum.

Well, Mr. Peale's crew had a terrible time, too. The banks of the pit kept caving in on them, and they got only a few bones for all their work. They gave up. Then they heard of bones in another farmer's swamp eleven miles away. They worked there with dozens of townspeople standing around the dig teasing them or cheering them on. Out of that mastodon burial, they found a set of ribs, some teeth, a few backbones, two tusks, a couple of toe bones, and the mastodon's broken shoulder blade.

Finally, on still another farm, they found some leg bones and part of the mastodon's head and lower jaw. It was enough to piece together a skeleton. They carved wooden teeth and bones to fill in those that were missing. It wasn't very scientifically done, but it was the first time a mastodon skeleton had been put together for people to see. It was an exciting exhibit, and it started the scientists of the day thinking about life on earth before man.

Now when someone like the celery farmer finds bones, all kinds of people work together to solve the mystery.

Men and women who specialized in things such as mammals, plants, the weather, glaciers, and pollen went to the farm with a bulldozer operator. As the summer went on and the dig grew, teachers and students who had come to see the dig helped search the fields for broken bits of bones.

The people on the mastodon hunt were armed with picks and shovels, trowels and whisk brooms, stakes and strings, surveyor's equipment, and the bulldozer. They didn't just start to dig holes all over the celery farm. That way they'd end up with bones but no data. The bones alone wouldn't mean much. They worked just like archeologists do.

The more they knew about a bone, the more valuable the bone became. The way a bone was placed next to another bone told them if the animal had died from a fall or a fight. The kinds of soil around the bone told the geologists whether the soil came from a lake bottom or a bog or a stream. The plant specialists searched the soil with microscopes to find grains of pollen that last for centuries. The pollen grains told them if the mastodon had lived in a pine forest or a maple woods. And the kinds of trees told the weather specialists what the climate was at the time.

It was an enormous detective story. Each specialist solved his clue questions, and when they were put together, they had the story of the mastodon from the celery field. With all the facts from the past, a zoologist added what he knew about how elephants live today, things such as how elephants in a herd help each other when they are hurt.

They knew from the size of the bones that the celery field mastodon was a bull. He had broken ribs that never quite

These mastodon bones were found by a farmer plowing his field. The long, curved bone in the back of the picture is one of the tusks.

healed. So they decided that this mastodon had been in a fight for leadership of his herd. If he had broken the ribs in a fall, the other mastodons would have helped him, and he probably wouldn't have died in a lake.

If he lost the fight, he was driven from the herd and wandered by himself for a while. He couldn't have died right after the fight because the ribs grew some at the ends. Probably in the early spring he wandered too far out onto the ice of the lake and broke through. If he had become stuck in the mud of a bog, the bones would have been buried closer together. If he had died on top of the ice, his bones would have been dragged to different places by the wolves and other flesh-eaters. The ankle, rib, and neck bones were buried close enough together to show that they were still held together by flesh and skin when they were buried.

All of it started when the celery farmer found the bones by accident. But the rest of the work was done by reading the scientific facts.

15

Protecting the Treasure

People do things in museums that seem hard to understand. They scribble on walls with lipstick. They color skeletons with crayons. They put gum on dinosaurs. They break in at night and steal. People who act that way don't care what happens to a museum's treasures. So the people who do care have to protect the treasures. In some museums dogs help.

Dogs are good museum guards, and all they cost is the daily dog food and someone to walk with in the park each day. German shepherds or Doberman pinschers are two of the best dogs trained for guard duty. They are strong and can be fierce when they need to be. A dog prowling the quiet halls at night can find a person who might have decided to hide behind an exhibit during the day. A dog can hold a thief against the wall and not attack until given a signal. It will bark until a watchman comes to help.

But dogs can't do everything. They can't answer people's questions or show visitors where things are. And if you let your dogs roam the halls during the day, some visitors might be frightened. You need other help.

You would probably hire guards, men and women who can

find lost purses and mothers of lost children. They can tell you how a mummy is made. But, like the dogs, guards can't be everywhere at once.

And you wouldn't want your museum to look like a bank anyway, with guards at every door. Nobody likes to see chains and locks and bars either. You would have to think of invisible ways to protect your treasures.

You'd think that keeping things behind glass would be protection enough. But it's not. Some very clever thieves know how to cut glass quickly and silently when no one is watching.

Some museums use small television cameras that can be almost invisible when they are built into a wall. The camera can be focused on a special exhibit with no guards in sight. In another room, a guard can watch the screens showing all the special exhibits and the people who walk by them.

Even if a museum can't afford to have a lot of television cameras, it can still guard valuable artifacts that are hanging on a wall or standing on a table.

What if a nice-looking old man decided that he just had to have an ancient Chinese mask for his living room. What if he went to the museum and waited until nobody else was in sight. Then what if he popped the mask into a bag. It might be days before a guard, making a check of the halls, would notice the empty space.

But what if you glued a special piece of metal to the back of the mask before you hung it up. And what if the man had to walk through an archway as he was leaving the building? The archway would be an electronic metal detector. It would be the same kind of detector used at airports to find people who might want to take a gun or bomb on the plane.

If the man walked through your detecting archway, a buzzer would sound and then the guards could politely check his packages. Many art galleries are using electronic detectors now. Just because museums show old-fashioned and ancient things, it doesn't mean they use old-fashioned methods of guarding.

16

Museum Kids

Where can you go to build a skeleton, fly a rocket, weave a poncho, make soap, feed a snake, dissect a shark, or hike to dig fossils? Where can you do all the things your family hates to have you do at home, like mush in clay or raise spiders?

A museum, of course. Museums need kids.

Museum kids are often the people who grow up to become curators and museum directors and taxidermists and geologists. Even when museum kids don't become scientists, they learn things at a museum that make them more interesting people anyway.

As you can tell by now, the whole job of a museum is education of one kind or another. It is a place where people learn, by studying the collections or just by looking at exhibits.

A long time ago, museum people decided that it was important to show kids the real things. Anyone can look at a movie of someone digging for fossils. Anyone can find a book to look at with pictures of fossils. But when you really find your own fossils and hold in your hand the remains of an animal that lived millions of years ago, you understand in a way that you will never forget.

Museum teachers love leftovers. In your museum you'd have leftovers, things too good to throw away but not quite good enough for the collections or exhibits. They might be duplicates, two of a kind. If you had two birds' nests, you might give one to the museum teachers. They might be things with no data. . . . Remember data? Those are the facts. An artifact that nobody knows anything about is no good for a collection, but museum kids can use it.

You can go to most museums after school or on Saturdays or in the summer to use these artifacts. You can find out how Indian women ground their corn or how Indian men chipped arrowheads. You can make a Chinese mask or cut the flesh off a dead animal to make a skeleton. You can learn how to make a coin collection or how to tell the Big Dipper from the Milky Way. In one way, a museum is like a school

Making a skeleton is messy work. This girl is learning how to do it.

because you learn there. But it's not like a school because you go there only when you feel like it and you learn what you like best.

What if some kid called you and said, "I can't get to your museum. We don't have a car, and my mom won't let me take the bus alone. But I really want to see your museum."

That would worry you, and you'd think, "H'mm, there must be lots of people who can't get to our museum building. Maybe we should take our museum to them."

There are many ways to take a museum to people.

You can make a satellite museum. Do you know what a satellite is? The moon is a satellite. It is a follower. It is part of our earth system, but it is separate, too. The big city museums send people to set up exhibits in empty stores or churches or schools. People can go right in their own neighborhood to see at least part of the museum. MUSE, the satellite of the Brooklyn Museum in New York City, was started in an empty store. It was a museum just for children. There were animals to touch and play with, and all kinds of good leftovers the kids could find out about. That satellite got so big and so many kids went there that they now have a big building of their own.

Lots of museums don't want kids running around because kids sometimes knock over exhibits or put sticky fingers on furry animals, even though they don't mean to hurt anything. But if the museum has a separate place with separate exhibits for kids to touch and play with, everyone can relax and have a good time. A funny thing happens then. The grown-up people start hanging around the children's museum because it's more fun.

Touching a dinosaur bone is better than seeing one in a picture. These bones are from the hind leg of a Brontosaurus that lived more than 300 million years ago.

Museum Kids

Even the museums that aren't for kids have begun to change their "look only" exhibits into exhibits that people can handle. Many places have "touch me" rooms, places where you can try on a pair of wooden shoes or knights' armor, where you can feel a skunk or an elephant's tusk, where you can light a fire with a flint or twist a milkweed stalk into a rope.

Some museums are sending out moving museums. Some of these are trains, some are buses, some are huge trucks, and some are just small delivery vans. But all of them carry exciting things you wouldn't see anywhere else.

Everyone is waiting to see whether a snake, a skeleton, or a skunk will be brought out of the Haul of Science, a van used by a museum to take science to the schools.

Color Wheels in Buffalo, New York, is an old school bus that has been made into a traveling art room. It visits playgrounds, and kids can paint or draw right there where the bus parks.

Artrain is a train, and kids who live near the tracks can see famous paintings or sculptures or paint things themselves. Many of the big trucks are called Museumobiles, and they take exhibits all over the cities.

Some smaller vans, like the Haul of Science in Buffalo, New York, don't have exhibits. But they carry baskets of things into classrooms for kids to see. On one day the Haul of Science might take Indian masks into one classroom, snakes into another, and hawks and owls into a third.

Museum education departments have classes for three-year-old children and eighty-year-old people, too.

17

What's Next?

Almost as soon as you think your museum is finished, it will be old-fashioned.

Science doesn't stand still. Every day there are discoveries, inventions, ideas, new ways of doing old things, new ways of building exhibits.

Now when you visit a museum, you walk past the exhibits. Maybe someday you will sit in a comfortable chair and all the exhibits will move by you on a huge conveyor belt.

Today in some museums you can push a button and smell the salty sea air of an ocean exhibit. You can push a button and hear the gulls scream or the wind screech. You can hear the monkeys howl in a jungle or the owl hoot in the freezing Arctic.

At the Boston Children's Museum, you can go into a manhole and crawl under the city streets to see the old trolley tracks, the water pipes, and the bundles of wires.

You can learn a tribal dance or try on a tribal costume at a People Center in the American Museum of Natural History.

At the Ontario Science Center, you can feel as though you are docking a space capsule, you can blow glass or look through a microscope. You can play games with a computer or stand in an exhibit and feel your hair stand on end as 300,000 volts of harmless static electricity pass over your body.

When you are old enough to work in a museum, what will museums be like? They could be nothing but ghost pictures.

A ghost picture is a new kind of photograph called a *hologram*. A hologram is a picture taken without a camera. It is taken by a laser, a very straight and powerful beam of light. There are many kinds of lasers. Some can burn holes in steel. Some can operate on your body without drawing blood. And some take pictures.

The hologram looks exactly like the real thing. It is not flat like the kind of picture we know.

A famous jewelry store in New York City had a display of diamonds. They used a hologram, a ghost picture, of a human hand sticking out of the glass holding a bunch of diamonds. It looked like a real hand and real diamonds.

Of course, everyone walking by said to himself, "Wow, what's this?" And they tried to grab hold of the handful of diamonds. Only there was nothing there. It was like a ghost. They could see it, but they couldn't touch it.

If you look on the back of a regular picture, you don't see the back of anything but the paper. But you can walk around a hologram and see all sides of it.

A museum could have a hologram of a mummy. It wouldn't be a real mummy, but it would look like one. You would think it was a mummy, but it would be only a picture that has three dimensions . . . that is, it is tall and wide and deep.

Museums change because people change. But no matter what new ways there will be to store information, museums will always have collections. Even if they show hologram ghost pictures instead of real artifacts, the artifacts will be in storage. Museums will always take care of the earth's treasures because that's what museums are for.

    What if you made a museum?

    Wouldn't that be great?

More Books to Read

TO FIND OUT ABOUT MUSEUMS AND THE THINGS THEY COLLECT

Museums

Burns, William. *Your Future in Museums.* New York: Richards Rosen Press, 1967.

Facklam, Margery. *Behind These Doors: Science Museum Makers.* Chicago: Rand McNally, 1968.

Katz, Herbert and Marjorie. *Museum Adventures.* New York: Coward, McCann and Geoghegan, 1969.

Just for Fun About Museums

Konigsburg, E. L. *From the Mixed-Up Files of Mrs. Basil E. Frankweiler.* New York: Atheneum, 1967.
This is a mystery story and fun to read even though it also tells how museums operate.

Insects

Conklin, Gladys. *The Bug Club Book: A Handbook for Young Bug Collectors.* New York: Holiday House, 1966.

Fossils, Rocks, and Dinosaurs

Andrews, Roy Chapman. *All About Dinosaurs.* New York: Random House, 1953.

Fenton, Carroll Lane. *Prehistoric World.* New York: John Day, 1954.
Holden, Raymond P. *Famous Fossil Finds.* New York: Dodd, Mead, 1966.
Hussey, Lois J., and Pessino, Catherine. *Collecting Small Fossils.* New York: Thomas Y. Crowell, 1970.
Williams, Henry Lionel. *Stories in Rocks.* New York: Holt, Rinehart and Winston, 1948.

Botany

Cosgrove, Margaret. *Plants in Time: Their History and Mystery.* New York: Dodd, Mead, 1967.

Astronomy

Branley, Franklyn M. *Experiments in Sky Watching.* Rev. ed. New York: Thomas Y. Crowell, 1967.
Gringhuis, Dirk. *Stars on the Ceiling: The Planetarium Story.* Des Moines: Meredith Corporation, 1967.
Reed, W. Maxwell. *The Stars for Sam.* Rev. ed. Edited by Paul F. Brandwein. New York: Harcourt Brace Jovanovich, 1960.

People

Clark, James. *Carl Ethan Akeley: In the Steps of the Great American Museum Collector.* Philadelphia: J. B. Lippincott, 1968.

Archeology

Benjamin, Nora. *The First Book of Archeology.* New York: Franklin Watts, 1957.
Edel, May. *The Story of Our Ancestors.* Boston: Little, Brown, 1955.
Friedman, Estelle. *Digging into Yesterday: The Discovery of Ancient Civilization.* New York: G. P. Putnam's Sons, 1958.
Mead, Margaret. *People and Places.* Cleveland: Collins-World, 1959.
White, Anne Terry. *All About Archaeology.* New York: Random House, 1959.

Index

Akeley, Carl, 53–55, 72–76
Albany, N.Y., 88
Albino, 49
American Museum of Natural History, 53
   African Hall, 71
   People Center, 105
   Stegosaurus, 84
Andrews, Roy Chapman, 78, 80
Anning, Mary, 77, 87
Anthropology, 17
Archeologists, 17–20, 22
Arrowheads, 4, 17, 20–21
Astrologers, 38
Astronomers, 38–42
Astronomy, 37, 38, 40, 41
   curator of, 37, 41
Artifacts, 18, 21–22
   protecting, 80, 107

Barnum, P. T., 53
Beetles, 30
Big Dipper, 40
Biological control, 34

Birds
   banding of, 43, 46
   that eat mosquitoes, 3
   eggs of, 5, 46–47
   flight of, 3
   freeze-drying of, 60–61
   study skins of, 44–45
Black holes, 40
Boston Children's Museum, 105
Boston News Letter, 89
Botanists, 32, 47
Botany, 25
   curator of, 10, 25–28
Bounty hunters, 47
Brooklyn Museum, 101
Buffalo, N.Y., 48, 104

Cabinets of curiosity, 7
Caribou, 48
Cockroaches, 29–30, 31
Collecting
   data, 27
   dinosaur bones, 78
   leaves, 26

Collecting (*continued*)
   live insects, 32–33
   meteorites, 41
   rocks, 12
Curators, 10–14, 63
   of anthropology, 17
   of archeology, 24
   of astronomy, 37, 41
   of botany, 10, 25–28
   of entomology, 29–33
   of geology, 10, 11–12, 16, 24
   of zoology, 10, 11, 43, 49, 51, 61

Data, 27, 91, 100
Decoys, 32
DDT, 34, 46–47
Digs, 90–91
Dinosaurs
   assembling, 81–83
   digging for, 78
   fake, 83–85
   finding, 77

Ecology, 4, 48
Education at museums, 99
Eggs
   of birds, 5
   of dinosaurs, 78–80
   of insects, 35
Electric detector, 96–97
Entomology, 29–30
Exhibits, 66–70, 71
Expeditions, 14
   Gobi Desert, 78–80
   Mount Mikeno, 72

Fiberglass, 58
Freeze-dried food, 59

Freeze-drying, 59–61
Frogs
   common green, 49
   orange, 48, 49

Gobi Desert, 78
Gorillas, 71–74
Guards
   dogs, 95
   men and women, 96
   television cameras, 96
Gypsy moth, 32

Herbariums, 26–27
Holograms, 106–107
Hornaday, William 52–53

Ichthyosaurus, 77
Insect(s)
   collections of, 31–32
   decoys for capturing, 32
   dermestids, 28
   killing jars, 33
   live collections of, 32–33
   museum bugs, 28, 45–46
   nets for capturing, 32
   research, 34–35
   setting board, 33
   sugaring for, 32–33

Jesuit missionaries, 19–20
Jumbo the elephant, 53–56

Killing jars, 33

Labels, 8–10, 34, 64
Lasers, 106

Leakey, Dr. Louis, 19
Leftovers in museums, 100–101
Light-years, 37, 42
Lucas, Frederic, 53

Mammoths, woolly, 89
Mastodons, 88–93
Matrix, 80–81
Mealworms, 46
Medicine Bow, Wyoming, 85
Meteorites, 42
Milky Way, 40
Mounds, 20
Moving museums, 103
    Artrain, 104
    Color Wheels, 104
    Haul of Science, 104
    Museumobiles, 104
MUSE, 101
Museum bugs, 28, 45–46
Museum kids, 99
Museum teachers, 100

Nature sanctuary, 48–49
Nets, insect, 32
Noah
    and the flood, 89
    and the raven, 90

Observatories, 38
Olduvai Gorge, 19
Ontario Science Center, 105
Orion, 40

Paleontologists, 78–80
Peale, Charles, 90
Periplaneta americana, 29
Pigment, 49

Planetariums, 40–41
Poison control center, 25
Pompeii, 22
Pottery, 4, 8–9
Preparators, 68

Rana clamitans, 49
Rattlesnake, frozen, 61
Registrar, 8–9
Rocks, 3, 10–12
Rodents, 47

Satellite museums, 101
Shale Creek, 48
Shrunken heads, 4, 8, 10, 17
Sites, 19–20
Skunks, 11–12
Skeletons, 46, 56
    assembling, 81
    dinosaur, 77, 81
    learning to make, 100
    mastodon, 88–90
Smithsonian Institution, 7, 8, 58
Space flights, 4, 41
Spearheads, 13–14
Stars, 4, 37–38
Stegosaurus, 38
Study skins, 44–45, 57
Sunspots, 38
Sugaring, 32–33

Tartars, 88
Taxidermists, 51–61, 65
Taxidermy, 51, 56
Telescope
    solar, 38
    radio, 40

Tom Thumb, 53–54
Trench, 20

United States Fish and Wildlife Service, 43
Universe, 40–41

Vacuum chamber, 61

Ward, Henry, 53
Ward's Natural Science Establishment, 52–55
Whale, blue, 58
Wolves, 47–48

Xanthochroism, 49

Zoologists, 43, 47–49